Mirror Exercises

Macro-Dimension Laboratory Series

Mirror Exercises

Macro-Dimension Laboratory Series

Claude Needham PhD.

Gateways Books and Tapes, Nevada City, California

ISBN (Trade Paperback) 978-0-89556-181-7
ISBN (PDF) 978-0-89556-600-3
ISBN (MOBI) 978-0-89556-602-7
ISBN (KINDLE) 978-0-89556-603-4
ISBN (EPUB) 978-0-89556-604-1

Cover design by Marvette Kort

Published by GATEWAYS / IDHHB, INC.
P.O. Box 370, Nevada City, CA 95959
(800) 869-0658; (530) 271-2239
http://www.gatewaysbooksandtapes.com

Names: Needham, Claude, 1951- author.
Title: Mirror exercises / Claude Needham.
Description: Nevada City, California : Gateways Books and Tapes, [2018] | Series: Macro-dimension laboratory series | Includes bibliographical references.
Identifiers: LCCN 2017031108 (print) | LCCN 2017043403 (ebook) | ISBN
9780895566003 (PDF edition) | ISBN 9780895566034 (Kindle) | ISBN 9780895566027 (MobiPocket eBook) | ISBN 9780895566041 (EPUB) | ISBN 9780895561817 (trade pbk.)
Subjects: LCSH: Observation (Psychology) | Mirrors.
Classification: LCC BF76.6.O27 (ebook) | LCC BF76.6.O27 N44 2018 (print) |
DDC 155.2--dc23
LC record available at https://lccn.loc.gov/2017031108

Table of Contents

Dedication

This book is dedicated to those that believe there is something happening here – even though what it is may not be exactly clear. However, after even a little exploration it is more than apparent that the doctrines espoused by the various religions, cults, and new age gurus are little more than half-guesses and outright self-serving propaganda.

For those folks that have intuited the glimmer of a truth peeking out between the cracks in reality we have assembled a rather large collection of mirror exercises. These experiments offer the opportunity to make personal observations personally observed.

There is nothing like personal experience, personally experienced. From personal experience you have the possibility of making your own observations, drawing your own conclusions, following your own path of investigation.

A small confession: between the covers of this book you will find a multitude of exercises that will inevitably lead the reader/doer to a multitude of observations which will in turn lead to a multitude of questions. That's not the confession. The confession is: you will find precious few (if any) answers in this book.

I suppose it is a bit unfair to provide such provocation.

If you are faint of heart and ill equipped for adventuring it might be best if you stopped reading and returned this book for a refund – assuming of course that you purchased the book.

This book is not for the timid or the mentally unstable. It is for the gentle and mentally fluid.

It has not escaped our attention that some of these exercises are potentially dangerous. They have the potential of providing transport out of the consensus reality into a magical world of what is – or what might be when the conditioning of human life is reduced.

Good journey, fair thee well.

In all likelihood we will not be around when you find this book in a trash heap in a corner of some burned-out city of some dystopian future.

I will leave you with advice that is appropriate for any wilderness traveler: be careful, don't take crazy risks, know your limits, be kind to yourself and your companions, look out for each other, and enjoy the journey – the horizon may not be the destination.

Introduction

Mirrors are magical devices – inspiring wonder and opening possibilities. Our own experience, literature, and media of all variety speak about mirrors as being much more than just sheets of reflective glass.

Alice uses a mirror as a portal into another world. Sleeping Beauty's step mother used a mirror to communicate with a genie. Merlin uses a mirror for divination and far-seeing.

Comedians will play with our funny bone by staging a scene in which two mimes simulate an individual standing in front of a mirror – by having mime mimic the actions of the other. Then finishing the skit by having one of the actors wave his hat or pinch the nose of the other. This is funny because deep down we are not fully convinced that the one we see in our own mirror is just a reflection.

How many books and movies use the situation of being trapped on one side of a mirror looking back into a world we cannot access?

In Harry Potter a mirror was used to show that which the viewer most desired, and also played a role in the plot by hiding a special object.

Any monster movie buff will recall that a classic vampire does not have a reflection in a mirror.

Once upon a time, the mythology about witches included a method of killing a witch by breaking a mirror while she was gazing into it.

Many mythologies include reference to seeing the true form of a shape-shifter when viewing them reflected in a mirror.

To avoid being turned to stone it was necessary to view Medusa through her reflection in a mirror (such as a polished shield).

The list goes on. There are countless references to mystical, magical, and downright weird properties of mirrors throughout literature. And why is this? Because glimpses of these experiences happen all the time. Granted the person on the other side of a mirror does not reach through and pull us in. However, the experience of that "other one" as alive, aware, and independent does happen from time to time.

If you subtract the theatrical melodrama from literary accounts of mirrors, you will be left with many good hints of what is possible. This should give you a good idea of where we are going with this book.

More about that later.

Organization of This Book

Once upon a time, several revisions ago, this book was organized into several chapters of introductory notes followed by a bunch of experiments.

This revealed itself as a very bad idea. There were so many notes and explanations that it took forever to get to the good stuff – the actual mirror experiments.

To fix this, we moved all the notes to the back of the book. This new book layout is better – but it is still not perfect.

Putting the notes at the back of the book does get you (the reader) into the mirror exercises without a ton of preliminary gibber-jabber. Unfortunately, some of that gibber and jabber happens to be notes important for the proper execution of the exercises.

Thus, we have the following suggestion. Take a small peek at the notes in the back of the book to gather an idea of what's there. This way, if you come across something in one of the experiments that requires further explanation, you can refer to the notes in the back for more in depth reading.

It might even be useful to glance at the notes periodically as you work your way through the book. In this way you increase your chances of stumbling upon helpful suggestions when you actually need them.

In case you haven't noticed, this is a book. Being a book brings about a very definite consequence – these experiments are listed one after the other. The sequential listing of the exercises is an unavoidable consequence of delivering content in book format. However, do not allow this accidental sequence fool you into believing that these exercises must be attempted in a strict linear fashion.

In the case of some exercises it is a good idea to do them in order – to help with gathering necessary *prerequisites*. However, it is equally useful to allow your intuition to play its part in the sequencing of exercises. So feel free to jump around and/or skip exercises as you feel inclined.

That said, let's get down to it. Next chapter begins the experiments.

Spider Push-Ups On A Mirror

Equipment:

Small flat mirror.

Prerequisite:

None

Place the mirror on a surface in front of you. This can be any level surface such as table, desktop or even your lap. The mirror is placed on its back, reflective side up. Position the mirror so that you may place your hand on it without straining either your arm or shoulder.

Place your hand on the mirror using your finger tips to create a five-legged spider-like form.

Adjust your view, so that you can see your hand along with its reflection in the mirror. If you cannot see the reflected spider, adjust your seating and/or the mirror position until you can see both your hand and its spider form reflection.

Have the five-legged hand-spider do push-ups on the surface of the mirror.

As the "spider" does its push-ups, you should see both the spider and its reflection doing push-ups in counter-point.

As you do the exercise, gather impressions.

You are not looking for any specific result. There is no

predictable TDM (ta-da moment). You may have a variety of cognitions, understandings, revelations, or moments of enlightenment. That's nice. But don't get hung up on it. This is not a race to some conclusive result. Nor is this a quiz.

As you do the exercise you will undoubtedly come upon various interesting variations. It is okay to check these out. Go ahead, follow these experiential pathways for short intervals. However, be sure to return to the initial simplicity of a five-legged hand-spider doing push-ups on a mirror.

This exercise can be repeated many times. Feel free to repeat as often as you wish.

Each time you do this exercise be sure to make an entry in your journal. This discipline will be useful in many different ways. Yes, you are expected to keep an experimental journal to capture the essence of your journey through these exercises and experiments.

Two Spiders Doing Push-Ups

Equipment:

No mirror required.

Prerequisite:

"Spider Push-Ups On A Mirror"

Have your left hand and right hand touch fingertips to fingertips – ring finger to ring finger, pinky to pinky, thumb to thumb, etc.

Place your hands in front of you at a comfortable distance and elevation.

Get the notion that between your hands is a mirrored surface, and that either your left hand (and/or your right hand) is a reflection of the other.

Holding this notion that one hand-spider is the reflection of the other, do spider push-ups between your hands.

Even though you *know* that you are using your left and right hands to create the double spider, experience it as if only one hand is yours and the other is its *reflection.* Continue doing spider push-ups in this fashion.

During this exercise it is best to have no specific goal. Gather impressions and be attentive to whatever comes your way.

Allow yourself to follow various experiential pathways for

short intervals, however, return to two spiders doing push-ups periodically.

If you feel so inclined, switch hands every now and then. Meaning, if the left hand was the reflection, have the right hand be the reflection. Or, if the right hand was the reflection, have the left hand be the reflection.

Be sure to recount the highlights of this experiment in your journal.

Spider Push-Ups On A Mirror #2

Equipment:

Small flat mirror.

Prerequisite:

"Two Spiders Doing Push-Ups"

Place the mirror on a surface in front of you.

Place your hand on the mirror using your fingertips to create a five-legged spider-like form.

When the mirror is positioned correctly, you should see your hand, looking a little like a spider on its five legs, along with its reflection in the mirror.

Have the five-legged hand-spider do push-ups on the surface of the mirror.

If this sounds all too familiar, it should. So far, this is the same as *Spider Push-Ups On A Mirror*.

Here, comes the difference.

As you do push-ups, image (i.e. hold the notion) that the spider hanging from the surface of the mirror is your hand, and that the spider standing on the mirror is the reflection.

If you feel so inclined switch hands periodically. Meaning, if the spider hanging from the surface of the mirror is your hand and the one standing on the mirror is the reflection, switch it up so that the spider standing on the mirror is your hand and the one hanging upside-down in the mirror is the reflection. For awhile have the spider on top of the mirror be your hand, then switch to the spider hanging upside-down in the mirror be your hand.

Be sure to recount the highlights of this experiment in your journal include any results, observations, impressions, or what-have-you.

Gazing In A Mirror

Equipment:

A floor mirror, wall mirror, or standing mirror.

Prerequisite:

None

Sit comfortably in a straight-backed chair.

Sit in front of the mirror so that you may easily gaze upon your reflection in the mirror.

Have your feet rest upon the ground, flat and comfortably apart.

Allow your hands to rest lightly in your lap, or upon the arms of the chair – assuming the chair has such.

Gaze, with diffused vision, at your reflection in the mirror. Use your face as the center of your field of view.

Allow everything other than the face in the mirror to be part of the wallpaper making up the environment.

Sit this way, gazing in the mirror for awhile.

When complete, do your journal entry for this experiment.

Two Hands Clapping

Equipment:

No mirror required.

Prerequisite:

None

Sit comfortably in a straight-backed chair.

Have your feet rest upon the ground, flat and comfortably apart.

Allow your hands to rest lightly in your lap, or upon the arms of your chair (assuming the chair has arms).

Allow yourself to relax and center.

When you feel that you have come into the present (i.e. you have separated *now* from the momentum of *then*) raise your hands so that they are about shoulder high in front of you.

Slowly clap your hands.

Continue clapping your hands.

Pay attention to the process. Gather impressions. Gather impressions as you are able. Gather impressions of this event in as much detail and depth as you are able.

After clapping your hands together for what seems like enough time, pause for a short while, then begin clapping again.

Slowly clap your hands for an additional amount of time.

Gather impressions. Gather impressions. Gather impressions.

Use the repetition of hand clapping to penetrate deeper and deeper into the experience of clapping your hands.

After clapping your hands together for this second session, break.

Pause for a moment to allow the reverberations of the experiment to settle.

Make a journal entry for this experiment.

[Note: Clapping in this instance has little to do with making the clap sound. We are not sitting in an audience expressing our appreciation to performers. When you were sitting in an audience expressing appreciation, then make an appropriately loud clapping sound. However, in this experiment, the clapping has more to do with the hands coming together and touching then making a sound.]

One Hand Clapping

Equipment:

Medium mirror.

Prerequisite:

"Two Hands Clapping"

Sit comfortably at a table or other surface upon which you may rest a flat mirror.

Place a medium-sized mirror flat on the table surface. Note: a medium-sized mirror is called for. This will be larger than the one used in the spider pushups.

Rest your hand (left or right) upon the surface of the mirror.

Raise your hand away from the mirror and then lower your hand back to the surface of the mirror. The combination of your hand and its reflection makes one hand clapping.

Do this one hand clapping for a while.

Gather impressions in as much depth and detail as you are able.

After clapping for what seems like enough time, pause for a bit then begin clapping again – this time using the alternate hand. If previously you were clapping the left hand onto the mirror, now use the right hand. If you were previously clapping the right hand onto the mirror, then use the left hand.

Continue clapping for this second session until it seems like enough, then break.

Pause for a moment to allow the reverberations of the experiment to settle.

As usual, be sure to journal.

Mirror Writing

Equipment:

A standing mirror.

Pad of paper.

Sharpie Pen (or its equivalent)

Prerequisite:

Nothing in particular

Arrange the pad of paper and the mirror so that you can see yourself writing by watching the reflection in the mirror.

Practice writing letters of the alphabet, numbers, and short phrases.

You could work with four lines from the *American Book of the Dead* if you like:

"All phenomena is illusion,

Neither attracted nor repelled,

Not making any sudden moves,

My habits will carry me through."

It is suggested that in addition to writing in your normal fashion, you alter your writing so that the words in the mirror are legible.

Masks In The Mirror

Equipment:

A standing mirror, or floor mirror.

Prerequisite:

Nothing in particular

While watching your reflection in the mirror, create a series of masks by making faces in the mirror.

Grimaces, grins, smiles, puckers, jaw twisters, tight-lipped, furrowed brow, smirk, scowl, surprise, et cetera, so on, and so forth.

Not a difficult exercise in concept. But it can be deceptively fruitful in execution.

Make notes in your journal after each session with this exercise.

Touch A Spot

Equipment:

A standing mirror.

Prerequisite:

Nothing in particular

Arrange the mirror on the table so that you can see a portion of the tabletop surface reflected in the mirror.

Select a spot on the table surface reflected in the mirror.

While looking in the mirror touch that spot on the table.

Select spots near, select spots far, select spots left, select spots right, basically select an ever changing variety of spots to touch.

Do this until you can touch the spot without hesitation.

Note: This may require several practice sessions before you can get to the stage of no hesitation.

Front And Back

Equipment:

A standing mirror.

An object – such as a book, a can, a bottle, a block, a ???

Prerequisite:

Nothing in particular.

Select an object to use for this experiment. The object can be a book, can, bottle, child's block, basically anything you have handy of that approximate size.

Arrange the object and the mirror upon the table so that you can see the backside of the object reflected in the mirror. This means you will be able to see the front of the object (sans mirror) in your normal field of view, as well as see the back of the object as it is reflected by the mirror.

Gaze at the object thus.

Make efforts to split your attention so that you may place your attention simultaneously on the front and the back of the object.

Portal

Equipment:

Medium mirror.

Prerequisite:

A few previous exercises

Sit comfortably in a straight-backed chair.

Sit so that you may easily gaze upon your reflection in the mirror.

Have your feet rest upon the ground, flat and comfortably apart.

Allow your hands to rest lightly in your lap, or upon the arms of the chair.

As you gaze into the mirror take onto yourself the notion that the mirror is a portal into another world.

Don't be fooled by the appearance of the mirror being a reflection of you and the room you are in. Look deeper into the reality of the mirror as a portal into another dimension.

Avoid

Equipment:

Medium mirror.

Prerequisite:

Nothing in particular.

Sit so that you are looking directly into the mirror – square on, as it were.

Sit two to three feet from the mirror.

Place your visual center of gravity in the middle of the mirror, but allow your visual focus to be diffused. [You can visit *Notes: Diffused Vision* at the back of this book if the concept of diffused vision is new to you.]

Avoid focusing your attention on the mirror.

This does not mean remove your view from the mirror. It means avoid placing your attention on the mirror.

Allow your attention to shift and fall anywhere you prefer – other than on the mirror.

Make a journal entry after completing the exercise.

Mine Eyes

Equipment:

Pretty much any type of mirror.

Prerequisite:

A few previous exercises

Sit comfortably in a straight-backed chair.

Sit so that you may easily gaze upon your reflection in the mirror.

Have your feet rest upon the ground, flat and comfortably apart.

Allow your hands to rest lightly in your lap, or upon the arms of the chair.

As you gaze into the mirror, look into the eyes of the one in the reflection. Ignore any and all phenomena. Just look into the eyes in the mirror.

As you look into the eyes in the mirror it is not uncommon to experience various emotions. You may experience the emotions as coming from the eyes in the mirror. Or, you may experience these emotions as your own reactions.

This is normal and not of any particular significance. Meaning don't get all psychological about it. Allow them to come and go without resistance – like watching ocean waves ebb and flow at the beach.

As you sit looking into the eyes reflected in the mirror you may find yourself catching glimpses of a deeper more constant emotion. Underneath all of the superficial emotions that come and go is a persistent emotion of deep love.

Allow the distracting emotional parade of cliché melodrama to pass by like driftwood floating down stream. Sometimes we like to tease ourselves with spooky visions or by getting all psychological. No need. Underneath it all is an emotion of deep and abiding love. Allow the superficial to drift on by.

Hold Attention

Equipment:

Medium mirror.

Prerequisite:

Nothing in particular.

Sit so that you are looking directly into the mirror – square on, as it were.

Sit two to three feet from the mirror.

Using diffused vision look into the mirror. [You can visit *Notes: Diffused Vision* at the back of this book if the concept of diffused vision is new to you.]

Hold your attention on/in the mirror without wavering.

The goal is to hold your attention unbroken on the mirror for a full sixty (60) seconds.

Whenever you notice a break in your attention, begin again.

Make the attempt ten (10) to fifteen (15) times.

Don't be surprised if it takes weeks or months before you are able to hold your attention unbroken on the mirror for a full sixty (60) seconds.

Make a journal entry after each attempt at this exercise.

Let The Space Build

Equipment:

Medium mirror.

Prerequisite:

"Hold Attention"

Sit so that you are looking directly into the mirror – square on, as it were.

Sit two to three feet from the mirror.

Place your attention on/in the mirror. As you hold your attention fixed on the mirror you will feel the space build.

You may experience this "building of the space" as a pressure, or perhaps a fullness, or perhaps like something else entirely.

You will need to find for yourself the indicators of the space building. All that is known for sure is that as you manage to hold your attention fixed without interruption there will be a *building of the space.*

Your goal is to notice when the space begins to build, and to notice when the space collapses following an interruption of your attention on the mirror.

This is not the only way in which a space may be built. It is one way – the way that you happen to be experimenting with at the moment.

Melting

Equipment:

Floor mirror, Medium mirror, Small mirror.

Prerequisite:

Some General Experience – i.e. a few previous exercises.

Sit comfortably in a straight-backed chair.

Sit so that you are looking directly into the mirror – square on – about two (2) to three (3) feet from the mirror.

Have your feet rest upon the ground, flat and comfortably apart.

Allow your hands to rest lightly in your lap, or upon the arms of the chair.

As you gaze into the mirror allow your attention to follow the sensations of melting.

At first the sensation may be very slight and uncertain. Follow the sensation and simply monitor its progress. Gradually allow the melting sensation to spread throughout your body.

Some folks find the mental image of sugar cubes melting in a class of mildly warm water useful.

After you gain facility with this exercise, experiment with following the melting sensations of the one in the mirror.

Zen Basics In Mirror

Equipment:

Standing mirror, Medium mirror, or Large mirror.

Zen Basics Kit

Prerequisite:

Touch A Spot.

Sit at the table as you would for any Zen Basics session.

Arrange the mirror and Zen Basic rocks on the table so that you can see the three Zen Basic rocks on the table *and* the three Zen Basics rocks reflected in the mirror – giving a total of six Zen Basics rocks.

Start the Zen Basics CD and run the standard process – working with six Zen Basics rocks – three on the table and three in the mirror.

NOTE: The Zen Basics CD is found in the "Home Kit" available on the ZenBasics.com website.

http://zenbasics.com/home-kit.html

Snap Shot/Screen Shot

Equipment:

Medium mirror.

Prerequisite:

Nothing in particular

Sit comfortably in a straight-backed chair.

Sit so that you are looking directly into the mirror – square on as it were – about two (2) to three (3) feet from the mirror.

Have your feet rest upon the ground, flat and comfortably apart.

Allow your hands to rest lightly in your lap, or upon the arms of the chair.

Gaze into the mirror.

As you sit, allow inner reverberations to quiet – like a pond coming to rest after a pebble has disturbed the surface.

When you feel the moment is right, figuratively press the Print Screen Button. In other words, take a snapshot of the space, the mirror, and the reflection in the mirror.

Back To Back

Equipment:

Floor mirror or Wall mirror.

Prerequisite:

Some General Experience – i.e. a few previous exercises.

Sit comfortably in a straight-backed chair.

Sit so that you are looking away from the mirror – with your back to the mirror.

Look back over your shoulder to be certain that you would be looking at yourself if you were turned around.

Have your feet rest upon the ground, flat and comfortably apart.

Allow your hands to rest lightly in your lap, or upon the arms of the chair.

Sit with your back to the mirror.

Be acutely aware of the reflection of your back in the mirror.

Experiment thus for awhile.

Make journal notes each time you try this experiment.

Mirror Practice

Equipment:

Medium mirror, Large mirror, or Floor Mirror.

Prerequisite:

Nothing in particular

To illustrate this exercise we shall assume for the moment that you have taken up contact juggling. Not a bad idea by the way. Contact juggling has many redeeming qualities and special surprises in store for you.

In any case, assuming that you have undertaken to learn contact juggling there will come a time when you will need to work on your isolations. An isolation involves moving your hand and body so that the ball appears to be suspended in space while you are moving around it.

This isolation can be something as simple as a doorknob type turn. Or, it can be something as challenging as the Palm Isolation.

In order to perfect these juggling moves you will need to practice in front of a mirror – watching in the mirror from the perspective of an audience member.

By the way, in addition to being a good practice to perfect one's juggling, it is also a good practice that can lead to very fruitful self-generated mirror exercises.

Normally when using a mirror to learn juggling, the time spent practicing in a mirror is aimed solely toward perfecting one's skill as a juggler. That is the entirety of one's focus.

For the purpose of the exercise "Mirror Practice" all that is required is to split your attention so that a portion is reserved for the task of noticing what is happening as you practice.

If you expand your "field of noticing" to include more than just those things related to your immediate goals you may stumble across hints of new exercises to explore with your mirror work.

Thus for this exercise we are suggesting that you practice a hobby, avocation, or skill in front of a mirror. Then pay attention. This can be playing guitar, contact juggling, French-Drop sleight of hand, drumming, crafts, et cetera.

As you work in this way, avenues of potential exploration will present themselves to you. As this occurs, take advantage of those opportunities of exploration.

Circus Parade

Equipment:

Medium mirror.

Prerequisite:

Nothing in particular

Sit comfortably in a straight-backed chair.

Sit in front of the mirror – about two (2) to three (3) feet from the mirror.

Rest your feet on the ground, flat and comfortably apart.

Allow your hands to rest lightly in your lap, or upon the arms of the chair.

In the circus there is a parade at the beginning of the evening. In the parade all of the performers (acrobats, clowns, trapeze artists, jugglers, lion tamer, horse back riders) circle the ring giving the crowd a peek at who and what's to come. As you gaze into the mirror watch without control the parade of characters across your face.

Waiting

Equipment:

Medium mirror.

Prerequisite:

Nothing in particular.

Sit comfortably in a straight-backed chair.

Sit in front of the mirror – about two (2) to three (3) feet from the mirror.

Have your feet rest upon the ground, flat and comfortably apart.

Allow your hands to rest lightly in your lap, or upon the arms of the chair.

As you gaze into the mirror, watch and wait.

Be attentive and vigilant without preconceived expectations. You know you are waiting for something. But, you don't know what it is. You only hope you recognize it when it comes.

Look Out Window

Equipment:

Medium mirror.

Prerequisite:

"Portal"

In "Portal" the idea was to consider the mirror as a portal into another dimension. In this exercise, consider the mirror to be a window – a window like any other window.

What does a window do? A window provides a look into the "outside". The outside is the stuff just on the other side of the glass. This is the place you would go if you should happen to walk through the door.

The difference between "Look Out Window" and "Portal" is that whatever you see through the window of the mirror is not in some dimension far, far away. It is right outside the door.

Outside Looking Back

Equipment:

Medium mirror.

Prerequisite:

"Look Out Window"

Sit comfortably in a straight-backed chair.

Sit in front of the mirror – about two (2) to three (3) feet from the mirror.

Rest your feet upon the ground, flat and comfortably apart.

Allow your hands to rest lightly in your lap, or upon the arms of the chair.

As you look into the mirror, get the very definite notion that you are outside, and that you are looking back into the room.

In this exercise you are outside looking in – not inside looking out.

Notice the difference.

Many Windows

Equipment:

Medium mirror.

Prerequisite:

"Let The Space Build" and "Look Out Window"

Sit so that you are looking directly into the mirror – square on, as it were.

Sit two to three feet from the mirror.

Begin by looking out the window. Consider the mirror to be a window and look out onto the exterior landscape.

Place your attention on the mirror and the landscape through the mirror. As you hold your attention thus you will feel the space build.

When the inevitable comes and you experience a break in attention, don't worry. Just start again. However, as you begin the process of looking out the window and holding your attention, be keenly aware that this is a different window. You are now looking out onto a different landscape.

Each time you have a break in attention and begin again you are looking out a different window.

Confess

Equipment:

Medium mirror.

Prerequisite:

A good number of the other exercises.

To confess (for the purposes of this exercise) is to tell it like it is – not tell it like it was.

When we hear the word confess, we automatically assume that we are expected to tell someone (priest, lover, interrogator, inquisitor, friend, therapy group, or psychiatrist) those immoral or illegal acts we have performed. Perish the thought.

Now is not the time, nor is it the place, to get into a full discussion of the nature of whose life it is anyway. Let it simply be stated: past events, actions, thoughts, and deeds are not the subject of this exercise.

We are after something much more profound and difficult to master for most humans – we are diving into the world of what is, not what was.

As you sit gazing into a mirror, confess to the other in the mirror what is.

Just say it like it is.

Keeping the focus on the present will not be easy, but it is most important that you do this.

For example, if you were to confess to the mirror self: “I have never done this before” that would not be in the present.

If you are feeling some kind of embarrassment sitting in front of a mirror talking to yourself, then say that: “I feel embarrassed sitting in front of a mirror talking to myself.”

If you suddenly notice a splash of light falling across the top of your dresser, then say that: “I see a splash of light falling across the top of the dresser.”

Keep your confessions of *what is* to the present. This means no analysis or emotional dialog.

“I am feeling embarrassed” is not actually fully in the present. “My face is flushed,” “my heart is beating a little faster than usual,” and “my mouth is a bit dry” are all in the present. When these symptoms are put together and run through the brain they become a calculation – which is at least one step removed from *what is*, and is actually a part of the past.

Not to worry if you find it a little difficult in the beginning to make the distinction. As you practice, you will naturally find yourself coming more and more into the present.

I Am Being Reflected In A Mirror

Equipment:

Floor mirror.

Prerequisite:

"Gazing In A Mirror"

A few background notes:

Let's consider for a moment how a scientist studying optics would set up these experiments.

A scientist would identify and define the four fundamental elements in each exercise.

1. The object.
2. The mirror.
3. The light.
4. The reflection of the object.

Those are the four key elements in each experiment.

Well, quite simply, in this exercise you are the object – the thing being reflected.

Sit comfortably in a straight-backed chair.

Sit in front of a floor mirror (preferred, or another mirror as dictated by availability).

Sit so that you may easily gaze upon your reflection in the mirror.

Have your feet rest upon the ground, flat and comfortably apart.

Allow your hands to rest lightly in your lap, or upon the arms of your chair (assuming the chair has arms).

Gaze, with diffused vision, at your reflection in the mirror.

As you gaze at your reflection in the mirror remind yourself periodically: "I am being reflected in a mirror."

Mull this over. Roll it around in your head. Consider it in all its aspects. "I am being reflected in a mirror."

Be sure to make an entry or two in your journal.

I Am The Reflection

Equipment:

Floor mirror.

Prerequisite:

"I Am Being Reflected In A Mirror"

Sit comfortably in a straight-backed chair.

Sit in front of a floor mirror (preferred, or another mirror as dictated by availability).

Sit so that you may easily gaze upon your reflection in the mirror.

Have your feet rest upon the ground, flat and comfortably apart.

Allow your hands to rest lightly in your lap, or upon the arms of your chair (assuming the chair has arms).

Gaze, with diffused vision, at your reflection in the mirror.

Recall the four key elements of a mirror experiment: the object, the mirror, the light, and the reflection of the object.

The traditional view is that you are the object. That is most definitely the traditional view. "Of course, I am the one here, looking into the mirror, looking at my reflection. The one in the mirror is a reflection of me that is here looking into the mirror."

For the purposes of this exercise, as you gaze upon yourself in the mirror assume the notion that you are the reflection.

In other words, you are not the object which is being reflected, you are the reflection of the object. You are the reflection. Have this notion that what you see in the mirror is not the reflection – it is the real you.

Some notes about "notions":

The American Heritage Dictionary defines a notion as:

1. A belief or opinion.
2. A mental image or representation; an idea or conception.
3. A fanciful impulse; a whim.
4. **notions** Small lightweight items for household use, such as needles, buttons, and thread.

If you blend all of these definitions into one big linguistic smoothie we should be getting close to the functionality we are aiming at.

Typically we are taught by the age of five that notions are not things which we should control. Ideas and beliefs are supposed to come from an authority for the consensus reality.

The idea that notions are changeable and that folks can assume onto themselves self-constructed notions is threatening to those who wish to control the status quo.

When you begin to use notions as a tool, the same way you might use a needle and thread, then all manner of possibilities open before you.

A notion does not derive its power from adherence to established consensus reality. A notion has power in its ability to help us establish set and setting for our meditations.

When you begin to take control over the assumption and renunciation of notions it is like finding the tuner on a car stereo or the satellite remote control. No longer are you subject to the content which is feed to you by coincidence or design of others.

Be sure to make an entry or two in your journal.

I Am Definitely Not The Reflection

Equipment:

Floor mirror.

Prerequisite:

"I Am The Reflection"

Sit comfortably in a straight-backed chair.

Sit in front of a floor mirror (preferred, or another mirror as dictated by availability).

Sit so that you may gaze upon your reflection in the mirror.

Have your feet rest upon the ground, flat and comfortably apart.

Allow your hands to rest lightly in your lap, or upon the arms of your chair (assuming the chair has arms).

Gaze, with diffused vision, at your reflection in the mirror. Use your face in the mirror as a centrum of viewing. Remember to keep your vision diffused.

As you gaze upon yourself in the mirror do your best to assume the notion that you are not the reflection.

Bring the full force of your will and attention to the task of pushing away all notions that you are the reflection.

Be absolute, adamant, and unyielding in your insistence that you are not the reflection.

Be steadfast in your resolve that your are the “real” one.

“That is the reflection, not me.”

“I am the real me, that is but a reflection.”

Work with this for twenty (20) to thirty (30) minutes. Or, if you get *cooking*, keep going for whatever time you prefer to set for yourself.

When you are done, break from the exercise and pause for a moment to allow the reverberations of the experiment to settle.

Be sure to make an entry or two in your journal.

Place, Put, Plop

Equipment:

Floor mirror.

Prerequisite:

"I Am The Reflection"

"I Am Not The Reflection"

On the next page, following these instructions, you will find a list of words. For this exercise you are asked to apply the instructions below to some, any, or all of the words in that list.

In the instructions below you will find the word "place." When running this exercise for each of the words in the list, substitute the new word into the instructions wherever the word "place" currently appears.

Start by running the instructions using the word "place", then repeat the instructions substituting each of the words in the partial word list.

As you gaze upon your reflection in the mirror:

Place your attention upon the mirror itself.

Place your attention upon your reflection.

Place your attention upon you sitting in front of the mirror.

Run these instructions ten to twenty (10 to 20) times for each word.

As you may notice there are quite a few words in the following list. Not a problem. Do a few words a day and you will be finished with the list in a week or two.

Please, do not feel it is necessary to complete this word list before moving on to other exercises within this book.

It can actually be quite useful to take this word list in bits and pieces. Working with this list over the course of the next month may give you the best context for penetrating this exercise and processing the results.

Word List:

Place, put, plop, deposit, distribute, establish, focus, plant, park, smear, accumulate, arrange, assign, bestow, bring, concentrate, affix, insert, lay, locate, paint, position, rest, rivet, set, settle, spread.

Place, Put, Plop, Thesaurus

Equipment:

Floor mirror.

Prerequisite:

"Place, Put, Plop"

After running each of the words found in the Place, Put, Plop exercise, you may wish to look up various words from that list in a thesaurus.

Look for additional words which grab your interest, attention, or simply catch your eye.

For each new word you find, run the previous exercise.

"New Word Here" your attention upon the mirror itself.

"New Word Here" your attention upon your reflection.

"New Word Here" your attention upon you sitting in front of the mirror.

Spot A Spot

Equipment:

Floor mirror.

Prerequisite:

"Gazing In A Mirror"

Sit gazing into a mirror.

Verbally give yourself the instruction "Spot a spot."

After you give yourself the instruction, carry out the instruction and spot (as in locate) a spot (as in a location).

After you spot a spot, verbally acknowledge your willing compliance with the request by saying one of the following: "Okay" or "Thank you" or "Good" or "Alright."

The reason for using a variety of acknowledgments is to avoid unnecessary roboticism.

The spot you spot may be anywhere within the room or within the reflection of the room.

It may seem very odd to be giving yourself verbal instructions, then carrying them out, then going the extra step of acknowledging yourself. That's okay. Do it anyway.

The prime directive in undertaking this type of exercise is "don't hurt self, don't hurt others." Odd, is not a problem.

Harmful is a problem. Whether it be harmful to you or harmful to another – don't do it. It is your responsibility as an experimenter to act from the center of your own integrity.

A little odd is not harmful – challenging to one's ego perhaps, but not actually harmful. If you can't tell the difference, it might be best to set this book aside. It is probably not for you.

Glad To Be Of Service

Equipment:

A floor mirror.

Prerequisite:

"Gazing In A Mirror"

Sit comfortably in front of a floor mirror so that you can gaze upon your reflection in the mirror.

Gaze, with diffused vision, at your reflection in the mirror – using your face as a centrum of viewing. Remember to keep your vision diffused.

As you gaze at your reflection, consider the notion that your presence here and now is supportive of the one in the mirror. Somehow and in some way which you may, or may not, understand your being there (sitting in front of the mirror) is of service to the one in the mirror. Take that as a given for the purposes of this exercise.

Beginning from that point of certainty that your presence is of service to the one in the mirror, work on maintaining the attitude "Glad to be of service."

As you gaze at the one in the mirror, infuse yourself with the attitude "glad to be of service." Let this attitude radiate throughout your presence like a perfume. Bask in this warm glow.

Make notes in your journal after completing this exercise.

Where Is The Reflection?

Equipment:

Medium, large, or floor mirror.

Prerequisite:

A good number of the previous exercises

While looking in, at, and through a mirror meditate on the question: “Where is the reflection?”

“I can see the reflection. I know, more or less, what is being reflected. I even have some idea about how reflections happen. But **where** is the reflection?”

Hold this koan – mull it over, work it.

Be A Mirror

Equipment:

No mirror.

Prerequisite:

"I Am The Reflection"

A reminder:

Previously (in the discussion of "I Am The Reflection") we discussed how a scientist studying optics would set up these experiments. We determined that such a scientist would establish four fundamental elements in the exercise.

1. The object.
2. The mirror.
3. The light.
4. The reflection of the object.

We have already done an exercise related to being the object. We have also done an exercise in which you worked with the notion that you were the reflection of the object. Now, as you may guess from the title of this exercise, you will work with the notion that you are the mirror.

Sit comfortably in a straight-backed chair.

Have your feet rest upon the ground, flat and comfortably

apart.

Allow your hands to rest lightly in your lap, or upon the arms of your chair (assuming the chair has arms).

Gaze, with diffused vision, at the whole of the room that lies in front of you. Relax your facial mask and let all tension drain from your face and scalp. Remember to not focus on anything in particular.

As you sit thus, be a mirror.

In front of you is the room.

On your surface is the reflection of the room.

Behind you is the icy emptiness of the void.

A mirror is implacable and unchanged by that which it reflects. The reflection is not on the mirror, of the mirror, in the mirror, past the mirror, by the mirror, through the mirror, or in any other way an aspect of the mirror.

A mirror is.

Everything seen in, around and about a mirror is a by-product of the mirror's ability to reflect light. Light coming to a mirror is reflected. That's it.

Be a mirror.

This does not mean to mimic what you see. This is not a psychological game of repeating back what you hear. Nor, does this involve any form of feedback.

Do not take the expression "be a mirror" as some form of New Age philosophy or psychology.

Be a mirror in the same sense that a sheet of glass with a coating of silver paint is a mirror.

Be sure to make an entry in your journal after you finish the exercise.

Mirror In A Mirror

Equipment:

A floor mirror.

Prerequisite:

"Be A Mirror"

Sit comfortably in front of a floor mirror so that you can gaze upon your reflection in the mirror.

Gaze, with diffused vision, at your reflection in the mirror. Use your face as a centrum of viewing – the center of your field of view. Remember to not focus on anything in particular.

As you gaze at the mirror, be a mirror as well.

What we have here is a mirror looking into a mirror – a reflection of a reflection reflecting a reflection.

I Am Loved

Equipment:

Medium, large, or floor mirror.

Prerequisite:

A good number of the previous exercises

Sit comfortably in a straight-backed chair.

Sit in front of a floor mirror (preferred, or another mirror as dictated by availability).

Sit so that you may gaze upon your reflection in the mirror.

Have your feet rest on the ground, flat and comfortably apart.

Allow your hands to rest lightly in your lap, or upon the arms of your chair (assuming the chair has arms).

Gaze, with diffused vision, at your reflection in the mirror. Use your face in the mirror as a centrum of viewing – but don't actually focus.

As you gaze upon the one in the mirror do your best to accept the fact that you are loved.

As thoughts arise which would deflect the notion that you are loved, gently dismiss them.

Accept without reservation that you are loved, unconditionally, by the one in the mirror.

You are not loved because of how you look, what you have done, or what you will do.

The one in the mirror loves without condition – and without emotion.

When you are done, break from the exercise and pause for a moment to allow the reverberations of the experiment to settle.

Be sure to make an entry in your journal.

Love The One In The Mirror

Equipment:

Medium, large, or floor mirror.

Prerequisite:

A good number of the previous exercises

Sit comfortably in a straight-backed chair.

Sit in front of a floor mirror (preferred, or another mirror as dictated by availability).

Sit so that you may gaze upon your reflection in the mirror.

Have your feet rest upon the ground, flat and comfortably apart. Allow your hands to rest in your lap, or upon the arms of your chair.

As you gaze upon the one in the mirror allow yourself to love that one without reservation. Notice and apperceive this unconditional love.

Love not because of how they look, what they have done, or what they will do. Love without condition – and without emotion.

When you are done, break from the exercise and pause for a moment to allow the reverberations of the experiment to settle.

Be sure to make an entry in your journal.

I Am Not This

Equipment:

Floor mirror.

Prerequisite:

"I Am Being Reflected In A Mirror"

"I Am The Reflection"

"I Am Not The Reflection"

As you may recall from "I Am Being Reflected In A Mirror" there are four fundamental components in the setup of a mirror experiment:

1. The object.
2. The mirror.
3. The light.
4. The reflection of the object.

As you gaze into a mirror at your reflection, affirm "I am not this."

Realize that you are not the object, you are not the mirror, you are not the light, you are not the reflection of the object. You are not this.

Affirm to yourself as a kind of mantrum: "I am not this."

Candle in the Mirror

Equipment:

Standing Mirror or Medium Mirror.

Candle

Prerequisite:

"Most Of The Above"

Place a candle so that you can easily view it in the mirror without strain.

Establish your standard mirror-gazing posture.

Light the candle.

Study the candle and candle flame in the mirror the way a lover studies the beloved.

This exercise/experiment can be expanded in so many different directions. I have written a book under the title: Candle Exercises as part of the Macro-Dimension Laboratory Series. Check it out.

Time Delay

Equipment:

Medium mirror.

An object such as a red ball.

Prerequisite:

Several of the preceding exercises

Arrange the mirror and red ball upon the table such so that you can see the red ball and the reflection of the red ball in the mirror.

When a pebble is dropped into a pool of water it takes a little time for the waves to travel from where the pebble hit the water to the edge of the pool. Light will travel much faster, but it still takes some time for it to travel from its source to its destination.

Because it takes this teeny tiny amount of time for light to travel, the ball you see sitting on the table and the ball you see reflected in the mirror are not from the same time.

The ball reflected in the mirror is from a time slightly older than the ball you see sitting on the table.

Split your attention to gaze at both the ball on the table and the ball reflected in the mirror while holding the notion that the two are not from the same time.

Living Hand

Equipment:

Medium, large, or floor mirror.

Your hand.

Prerequisite:

Several of the preceding exercises

Place the mirror on the table in front of you (slightly to the side) so that you can see your hand (left or right) reflected in the mirror without also seeing your face and body.

While watching your hand in the mirror, have it caper about on and off the table. Allow the hand to come alive and assume a volition of its own.

Two Hands Time Delay

Equipment:

Medium mirror.

Your Hand

Prerequisite:

Time Delay and Living Hand

Place the mirror on the table in front of you (slightly to the side) so that you can see your hand (left or right) reflected in the mirror without also seeing your face and body.

When a pebble is dropped into a pool of water it takes a small but finite amount of time for the waves to travel from where the splash to the edge of the pool. Light will travel much faster, but it still takes some small finite time for it to travel from its source to its destination.

Because of this tiny amount of time required for light to travel, the hand you see sitting on the table and the hand you see reflected in the mirror are not from the same time.

The hand reflected in the mirror is from a time slightly older than the hand you see sitting on the table.

Split your attention to gaze at both the hand on the table and the hand reflected in the mirror while holding the notion that the two are not from the same time.

Mirror As Center Of Universe

Equipment:

Small, medium, large, or floor mirror.

Prerequisite:

Nothing in particular

Focus your attention upon the mirror.

Make that mirror be the center of the universe. Consider that everything that exists and everything that has happened was part of a grand scheme in order to support that mirror being here and now.

The glass industry and the science behind the manufacture of glass arose in order to make the glass out of which your mirror was fashioned.

The paint industry and chemistry itself was evolved in order to create the painted backing of the glass so that it would become a mirror.

The transportation industry, trucks, Detroit, truck drivers, the Department of Motor Vehicles, and road construction crews were developed to allow for the mirror being moved from where it was manufactured to your room at this moment.

Et cetera, so on and so forth.

Consider how each and everything which exists today is part

of the elaborate fabric of reality required to support the existence of that mirror.

Mirror In The Dark

Equipment:

A floor mirror.

Prerequisite:

Several of the proceeding exercises.

Sit comfortably in front of a floor mirror so that you can gaze upon your reflection in the mirror.

Gaze, with diffused vision, at your reflection in the mirror. Use your face as a centrum of viewing, without actually focusing.

Sit this way for a few minutes until you have established a connection with the reflection in the mirror.

Close your eyes.

Sit, eyes closed, maintaining the connection with the figure in the mirror.

When you lose connection with the one in the mirror, open your eyes and re-establish connection, then close your eyes again.

[Note: It is very possible to re-establish connection to the mirror self without opening your eyes.]

Speak To The Mirror

Equipment:

A floor mirror.

Prerequisite:

Several of the proceeding exercises.

Sit comfortably in front of a floor mirror so that you can gaze upon your reflection in the mirror.

Gaze, with diffused vision, at your reflection in the mirror. Use your face as a centrum of viewing, without actually focusing.

Carry on a conversation with yourself, while looking at your reflection in the mirror.

Mirage

Equipment:

Medium mirror.

Prerequisite:

Several of the previous exercises

In contrast to a hallucination, a mirage is a real optical phenomenon which can be captured on camera, since light rays actually are refracted to form the false image at the observer's location. What the image appears to represent, however, is determined by the interpretation of the viewer.

Sit comfortably in front of the mirror so that you can gaze upon your reflection in the mirror.

As you look into the mirror, take on (assume) the notion that the reflection you see is just an optical phenomenon, an illusion, covering the true view of what would be there to see if only the reflection was not obscuring it.

Clean The Mirror

Equipment:

Floor mirror, or Large mirror
Glass cleaner
Cleaning cloth or rag

Prerequisite:

Nothing in particular

Stand in front of the mirror.

Using the glass cleaner and cleaning cloth, clean the surface of the mirror.

Notice the person in the mirror that is cleaning the other side of the mirror in perfect synchronization with you.

For every spot you clean, they are cleaning the corresponding spot on the other side of the glass. You clean a spot, they clean a spot right underneath.

Every time you spray more glass cleaner on the mirror, they spray more glass cleaner on the mirror from the other side.

Perfect synchronization.

Pretty amazing teamwork if you think about it.

Follow & Contribute

Equipment:

Floor mirror, or Large mirror

No glass cleaner

No cleaning rag

Prerequisite:

Clean The Mirror

Stand in front of the mirror.

Bring your hand up to the mirror as if you had a cleaning cloth and was about to wipe the surface.

Do not touch the mirror. Keep your hand about an inch from the mirror's surface.

Move your hand back and forth across the surface of the mirror (without touching).

As you do this, notice the one in the mirror following your every movement.

In doing this exercise/experiment there are several different sub-experiments that you can concentrate on.

1) Notice your hand. Place attention on your hand as it moves across the surface of the mirror (remember, no touching). Attend to your hand as if it were a moving sculpture. Notice its position, tilt, and movement through the space.

2) Sense the physical presence of the hand in the mirror that is following your hand.

3) Instead of having the hand in the mirror follow your hand, you follow the hand in the mirror as it moves.

4) When following the hand in the mirror, you may either drag slightly behind the other hand thus letting it pull you along as a willing but passive participant, or, you can contribute to the motion by anticipating the direction of movement and actively moving in that direction.

Try out these variations and any others that may occur to you as a result of exploration in this experiment.

Ebb & Flow

Equipment:

Floor mirror, Large mirror, or Medium mirror.

Prerequisite:

A good number of the other exercises.

Sit comfortably in straight-backed chair.

Have your feet rest on the floor, at a comfortable distance apart.

Allow your hands to rest lightly in your lap, or upon the arms of your chair.

Allow yourself to relax and center.

When you feel that you have come into the present, allow yourself to gently follow your breath. Do not change your breath. Just notice the in and out process as you breath.

When you breath in, sense your body filling with vibrant energy flowing in through the full surface of your body.

As you exhale, sense the ebb of this buzzing fluid out the bottom of your feet, across the distance between you and the one in the mirror, then up through the feet of your reflection – filling the one in the mirror with liquid light.

There is more than one way to view this process.

One option is to consider the collection as an accumulation of prana from the local environment which you then send through the mirror to the one sitting in the chair opposite you.

That is just one option. As you work with this exercise your perspective about what may (or may not) be occurring will change. Make note of any new perspectives about this exercise in your journal.

This is an exercise that can easily turn into a practice.

Sharing the Same Atmosphere

Equipment:

Floor mirror, Large mirror, or Medium mirror.

Prerequisite:

"Ebb & Flow"

Sit comfortably in straight-backed chair.

Have your feet rest on the floor, at a comfortable distance apart.

Allow your hands to rest lightly in your lap, or upon the arms of your chair.

Allow yourself to relax and center.

When you feel that you have come into the present allow yourself to gently follow your breath. Do not change your breath. Just notice the in and out process as you breath.

As you attend to your breath – in & out – make note of the fact that you and your reflection are sharing the same atmosphere.

When you and your reflection breath in, you both are drawing from the same pool.

When you and your reflection breathe out, you both exhale into the same pool of air.

Buoyant

Equipment:

Medium mirror.

Prerequisite:

At least a few of the previous exercises

Hopefully you have been on a raft, canoe, boat, inner tube, surf board, or some other flotation device. If not, then what I'm about to suggest will make no sense – since you will not have the personal experience (personally experienced) required to make the connection. If that is the case, then it is recommended that you gather a little more experience. One way to do this would be through the book *Tabletop Demos*, also (such a coincidence) written by the author of this book.

Sit comfortably in front of a floor mirror, or large hanging mirror, such that you can gaze upon your reflection in the mirror.

Gaze, with diffused vision, at your reflection in the mirror. Use your face as a centrum of viewing, without actually focusing – meaning, don't concentrate your visual attention upon the face to the exclusion of interest in the rest of the visual field.

As you gaze into the mirror, allow the sensation of floating buoyancy to proceed – a gentle rising and falling as if you were sitting in a raft upon the water.

Gazing implies looking without specific intent. This would be a good approach for watching the rising and falling sensation as well. Just look. There is nothing, in particular, to notice. There is nothing, in particular, to do. There is nothing, in particular, to enhance. There is nothing, in particular, to suppress. There is nothing, in particular, to succeed at. There is nothing, in particular, to fail to do. Just watch with your inner sensing the buoyant rising and falling as you continue to gaze into the mirror.

Looking Up

Equipment:

Medium mirror.

Prerequisite:

Some General Experience

Sit comfortably in a straight-backed chair.

Sit in front of the mirror – about two (2) to three (3) feet from the mirror.

Have your feet rest on the ground, flat and comfortably apart.

Allow your hands to rest lightly in your lap, or upon the arms of the chair.

As you gaze into a mirror, get the definite sensation that you are looking up into the mirror, as if you are laying on the floor looking up into a mirror attached to the ceiling.

Hanging Down

Equipment:

Medium mirror.

Prerequisite:

"Looking Up"

Sit comfortably in a straight-backed chair.

Sit in front of the mirror – about two (2) to three (3) feet from the mirror.

Have your feet rest on the ground, flat and comfortably apart.

Allow your hands to rest lightly in your lap, or upon the arms of the chair.

As you gaze into a mirror, get the definite sensation that you are suspended over the mirror looking down into it.

This would be a little like looking into a well, seeing your reflection in the well below you. Only in this case it is more like the room has turned on its side, as if your back is glued to the chair and you are being held suspended over the mirror.

True Face

Equipment:

Medium mirror, Large mirror, or Floor mirror..

Prerequisite:

Some General Experience

Sit comfortably in a straight-backed chair.

Sit in front of the mirror – about two (2) to three (3) feet from the mirror.

Have your feet rest upon the ground, flat and comfortably apart.

Allow your hands to rest lightly in your lap, or upon the arms of the chair.

As you gaze into a mirror, look for the true face underneath, behind, over, and in between the lines of the apparent face. Somewhere in there you'll find the true face. The face that is the most you. Look for that face.

The Gallery

Equipment:

Medium mirror, Large mirror, or Floor mirror..

Prerequisite:

Some General Experience

Sit comfortably in a straight-backed chair.

Sit in front of the mirror – about two (2) to three (3) feet from the mirror.

Have your feet rest upon the ground, flat and comfortably apart.

Allow your hands to rest lightly in your lap, or upon the arms of the chair.

As you gaze into the mirror, consider that above the ceiling is a gallery of individuals – larger than life and unseen by you. This gallery looks through the ceiling watching and observing you as you gaze into the mirror.

Adopt the notion that you are being observed by this gallery.

Magic In The Mirror

Equipment:

Large Mirror

DVD: Secrets of the French Drop by E.J. Gold

DVD: Magic in the Mirror by E.J. Gold

Prerequisite:

Most of the above exercises.

[Note: These are not the actual instructions for "Magic in the Mirror". This page is a placeholder and pointer to the existence of this ritual method. It is a very powerful practice that should not go unknown.]

Here for your edification are a few quotes about the DVD Magic in the Mirror:

"The instructions on this DVD are very specific. You are given everything you need to perform the rituals shown. If you are capable of learning the simplest and most basic magic trick on the planet (The French Drop) then you are capable of following the path illustrated in this DVD."

"Magic in the Mirror is something you must see. It is a thing of beauty and magic. It's a portal to our other selves and other worlds; ancient worlds and practices... It is the mystery of Being and Reality. Even if you have no intention to ever do the actual 'Magic in the Mirror' practice, you owe it to yourself to watch this DVD at least once."

Infinity Cube

Equipment:

Infinity Cube. (Special Equipment)

Prerequisite:

Most of the above exercises.

The Infinity Cube is a special box created with regular mirrors and "one way" mirror.

The inside dimensions of the box are optimal at 12" wide, 12" deep, and 12 1/8" height. Why? Because with these dimensions you can use inexpensive 1/8" mirror tiles for the bottom and three walls of the cube.

One wall will use "one way" mirror – with a large hole cut in the side wall. This allows a viewer to put their face up to the cube and peer through the "one way" mirror viewport into an infinite reflection of the cube and its contents.

What to do with your infinity cube?

Let me give you two more hints before you are left to your own devices:

1. If you leave the top of the cube open – thus allowing for a light source to illuminate the contents of the cube – then you will see the inner space of the cube reflected dozens or hundreds of times.
2. If you cover the top of the cube with an additional

> mirror – thus closing off the cube to any light source other than that coming in through the "one way" mirror viewport – then you will see your face peering in through the viewport reflected dozens or hundreds of times.

Okay, now you are on your own.

If you have done the exercises included in this book you should have a very good start for working with an Infinity Cube.

Crystal Cave

Equipment:

The Crystal Cave. (Special Equipment)

Prerequisite:

Most of the above exercises.

The Crystal Cave is an Infinity Cube made so large that a person can enter into the mirrored space – closing a mirrored door behind them – thus sitting in a space containing infinite reflections of themselves and whatever is in the space.

If you are one of the lucky few with your own full size Infinity Cube called The Crystal Cave, then use yours.

If you are like most of us, you will need to rent time in a commercial Crystal Cave.

The instructions for this exercise are simple.

Step 1) Get in the Crystal Cave.

Step 2) Be in the Crystal Cave.

Step 3) Get out of the Crystal Cave.

That's it.

We have deliberately not included any instructions on what to do in the Crystal Cave. And, we have deliberately not given any clue about what to expect from being in the Crystal Cave.

Unless you come to one of our retreats or advanced workshops, you are best not having any specific instructions.

After completing the bulk of the exercises in this book you should be well prepared to follow a thread of exploration as it presents itself to you in the Crystal Cave.

At a workshop or retreat we can work with you to follow specific threads. However, even then you will find us hesitant to limit your exploration by creating a narrow confine of instructions.

Yes, there are very specific effects and definite realizations which can come from work in the Crystal Cave. But how you get to these may not be the same way that your next-door neighbor gets to them.

Partner Work

The following exercises are designed to be performed with a partner. Your partner can be any pal, paramour, or buddy that has an affinity for this type of work.

Ideally, your partner will be doing the individual exercises right along with you. In this way you will both gain maximum benefit. However, it is possible that your partner may not be as inspired by these exercises as you are; and, he or she is simply joining in as a service to you. This is okay too.

Consider yourself lucky if you can find a partner willing to do these exercises with you. It is not often easy to find anyone willing to sit still for this type of shenanigans.

Gazing At A Partner

Equipment:

A partner.

No mirror required.

Prerequisite:

None

Sit comfortably across from a partner so that you can gaze at each other.

Gaze, with diffused vision, at your partner's face. Use his or her face as a centrum of viewing. Remember to keep your vision diffused.

Have it be that everything other than your partner's face is part of the tapestry from which the perceived environment is formed. Consider the only portal of living vibrant non-wallpaper to be your partner's face.

Journal your experiences with this experiment.

[Note: After doing this exercise revisit “Gazing In A Mirror.”]

Partner As My Reflection

Equipment:

A partner.

No mirror required.

Prerequisite:

"Gazing At A Partner"

You and a partner each sit comfortably in a chair so that you can gaze at each other.

As you gaze at your partner get the definite notion that you are looking into a mirror at your own reflection.

Admittedly, the reflection looking back at you may be a bit different than the one you would normally expect. Don't let that get in the way.

It if helps, you could hang an empty picture frame half way between you and your partner. Or, sit on opposite sides of a sliding glass door. This may help establish the "mirror" aspect of this exercise.

Do this for fifteen to twenty (15 to 20) minutes.

Make notes in your journal after you finish. By the way, you and your partner should not discuss this exercise at all until after you have finished your journal entry.

Glad To Be Of Service, Partner

Equipment:

A partner.

Prerequisite:

"Gazing At A Partner"

Sit comfortably across from a partner so that you can gaze at each other.

Gaze, with diffused vision, at your partner's face.

As you gaze at your partner, consider the notion that your presence here and now is supportive. Somehow (and in some way which you may, or may not, understand) your being here is of service to your partner. Take that as given for the purposes of this exercise.

Beginning from the perspective that your presence is of some service to your partner, work on maintaining the attitude "Glad to be of service."

As you gaze at your partner, infuse yourself with the attitude "glad to be of service." Let this attitude radiate throughout your presence like a perfume. Bask in this warm glow.

Be A Mirror – Partner

Equipment:

A partner.

Prerequisite:

"Gazing At A Partner"

"Where Is The Reflection"

Sit comfortably across from a partner so that you can gaze at each other.

Gaze, with diffused vision, at your partner's face.

As you gaze at your partner, assume the posture of a mirror.

A mirror is implacable and unchanged by that which it reflects. The reflection is not on the mirror, of the mirror, in the mirror, past the mirror, by the mirror, through the mirror, or in any other way an aspect of the mirror.

A mirror is.

Everything seen in, around and about a mirror is a by-product of the mirrors ability to reflect light. Light coming to a mirror is reflected. That's it.

Be a mirror.

This does not mean to mimic what you see. This is not a psychological game of repeating back what you hear. Nor does this involve any form of feedback.

Do not take the expression “be a mirror” as some form of New Age philosophy or psychology.

Be a mirror in the same sense that a sheet of glass with a coating of silver paint is a mirror.

Be sure to make an entry in your journal after you finish the exercise.

Two Mirrors

Equipment:

A partner or partners.

Prerequisite:

"Be A Mirror"

Sit comfortably side by side with your partner or partners.

Sit so that you are all facing the same direction.

As you gaze straight ahead, have it be that each of you is a mirror reflecting the same chamber.

Contemplate, consider, meditate, mull over, muse, ponder, ruminate, study on, weigh, wonder and reflect upon your mutual nature as mirrors.

Partner In The Dark

Equipment:

A partner.

Prerequisite:

"Gazing At A Partner"

Sit comfortably across from a partner so that you can gaze at each other.

Gaze, with diffused vision, at your partner's face. Use his or her face as a centrum of viewing, without actually focusing.

Sit this way for a few minutes until you have established a connection with the beingness of the other.

Close your eyes. Or, use a hood. Or, use a blindfold.

Sit in the presence of your partner.

Revisit: After you have done this exercise, revisit "Mirror In The Dark."

Partner In The Mirror

Equipment:

A partner.

A floor mirror.

Prerequisite:

"Gazing At A Partner"

Arrange you and your partner's seats so that you can each see the other reflected in the mirror – but not see yourselves.

Gaze, with diffused vision, at your partner's face reflected in the mirror. Use his or her face as a centrum of viewing, without actually focusing.

Sit this way for awhile.

Speak To The Mirror – Partner

Equipment:

A partner.

A floor mirror.

Prerequisite:

Several of the proceeding exercises.

Arrange you and your partner's seats so that you can each see the other reflected in the mirror – but not see yourselves.

Gaze, with diffused vision, at your partner's face reflected in the mirror. Use his or her face as a centrum of viewing, without actually focusing.

Carry on a conversation of any sort while looking at each other only as reflected in the mirror.

Revisit: After you have performed this exercise revisit “Speak To The Mirror”

Afterword

While it is true that we have included a fair number of mirror exercises and experiments in this book, it is far from comprehensive.

The vision for this book was to provide enough material to plant a participant's feet firmly on the path of mirror work. Hopefully this has been accomplished.

Most likely you – or a friend reading over your shoulder – will have a favorite exercise that happens to be missing from this volume.

Your favorite exercise could be missing from this book because of one of the following reasons:

1. It slipped our minds.
2. We felt it should only be attempted under supervision.
3. We didn't want to be responsible for it tunneling through to the 26th Century.
4. We felt it could become a misdirection or detour.
5. Or, some other reason we don't want to mention.

Of course there is yet one additional reason. There are many exercises which are easily suggested by those included in this book. So you could say they have been left for you to discover through the process of extrapolation.

Notes: Experiment, Exercise, Activity, Recipe, Meditation, and Practice

While it is true that in this book we refer to exercises and occasionally experiments, we view each of these activities are much more than simply experiments and exercises. They may also be viewed as activities, recipes, meditations and, on occasion, practices.

Experiment: Investigate in the classic fashion of science: observation, postulate, followed by further observation.

Exercise: Working to strengthen muscles, talents, and abilities.

Activity: Something that one does with no particular goal or reason – not task-directed. Something you do simply because you are doing it.

Recipe: a set of instructions for making or preparing something, a method to attain a desired end.

Meditation: To immerse one's self in a spiritual perspective and act from a spiritual perspective – self-initiated immersion.

Practice (or discipline): A regular or full-time performance of actions and activities undertaken for spiritual development or carrying out a spiritual purpose.

Each of the recipes contained within this book may be taken as an experiment, an exercise, an activity, and a meditation. It all depends upon where you are in the process.

While the above listing may appear at first glance to form a ladder, it would be better to consider four perspectives as a spiral.

Any given recipe may begin as an experiment, then pass through the stages of exercise, activity, then meditation only to reemerge as an experiment on a different level. This is as it should be.

Notes: Mirrors And Other Equipment

Within this book we shall make use of several different types and sizes of mirror.

Tiny mirror: This may be something on the order of a business card in size. We find signaling mirrors from your standard backpacking store to work well for this size. These are typically made of stainless steel. They may or may not have a hole in the center. The hole is not necessary. They are carried by campers with the idea in mind of using them to signal searchers in the advent of being lost. Compact cosmetic mirrors will also work for this size.

Small mirror: This would typically be the size of a hand mirror – maybe something in the 8 inch by 10 inch size. Or, if you happen to have a builder's supply near you, check out the mirror tiles. These come in 12 inch by 12 inch squares. This is not a bad size. However, they can be fragile and require special handling. This fragility issue can be partially fixed by framing or mounting the mirror tile.

Standing mirror: This can be a makeup mirror or a small mirror mounted in a frame with standing supports. A standing mirror is basically a small mirror which you can stand upright on a table. This mirror should be able to be either tilted or completely upright.

Medium mirror: This would be of the two foot by two foot variety. If you don't have access to a medium-sized mirror, sometimes it is possible to substitute a grid of small mirrors. Depends entirely on the experiment in question.

Floor mirror: A free-standing floor mirror is movable and can hopefully be repositioned at need. It would be handy if the mirror can be tilted. But this may not be a requirement.

Wall mirror: There may be experiments which require a full wall mirror. This would be a mirror such as found in a dance studio. If you can't afford such a mirror setup yourself, perhaps you can avail yourself of a nearby dance studio.

Notes: Mirrors Are Glass

While it it true that mirrors do not have to be made of glass, fact is that most mirrors are made of glass.

Glass breaks.

Broken glass is sharp.

Sharp bits of broken glass can cut flesh.

Your body is made of flesh and bone.

The shards of glass from a broken mirror can cause injury.

We expect you to take care, and be fully responsible anytime (and every time) you work with mirrors.

We have managed to do all of the experiments found in this book without breaking a single mirror nor cutting ourselves even once. We would hope that you can manage to display the same modest ability to work with a standard mirror without damage to either the mirror or yourself. If you manage to shave, apply makeup, check a hem, comb your hair, and do any number of tasks involving mirrors without accident, you should also be able to do these exercises without damage to yourself.

If you feel the challenge of looking in a mirror without breaking it is beyond you, please return this book to wherever you acquired it.

Notes: Experiment Prerequisites

The order in which you do the experiments in this book is not absolute. We have assembled the exercises in a sequence convenient for the layout of the book. However, that does not mean that you must do the exercises in the same linear sequence.

We have provided some indication of prerequisites when appropriate. Below is a brief description of the designations used.

"None" or "No Prerequisite": There are many experiments which have no prerequisite. For example: Spider Push-ups and One Hand Clapping. Neither has a prerequisite. This means you could do either one of these without previous experience.

Specific Prerequisite: There are experiments which have a specific prerequisite. For example the experiment "Two Spiders Doing Push-Ups" has the prerequisite "Spider Push-Ups On A Mirror". This is for a reason. Trust us, things will go better if you do the "Spider Push-Ups On A Mirror" before you do "Two Spiders Doing Push-Ups". That is why we suggest it as a prerequisite. Sometimes the prerequisite will itself have a prerequisite which may or may not have a prerequisite itself. This chain of prerequisites will form a sequence. Follow the sequence. Things will go better.

"Some General Experience": There are a few experiments which have no specific prerequisite, but, do have the prerequisite of a general background and some prior experience. These experiments generally require that you have performed several (three to seven) experiments before attempting them.

"Most Of The Above": These experiments are of the typc that you should have done a major portion of the experiments in this book before attempting them. "Crystal Cave" is one such experiment. These is nothing to prevent you from doing it early on in your work with mirrors. But, since the "Crystal Cave" can be expensive to do you owe it to yourself to gather significant preparation before attempting. Give yourself every possibility of having your expense and your effort well rewarded.

Notes: Relatively Safe Environment

When involved in any type of real experimentation, it is not possible to create a 100% safe environment. In fact, given the propensity for meteorites, lightning, floods, drive-by shootings, mad cow disease, crazed neighbors, bird flu, Nile virus, or any number of real dangers, don't you think the notion of a 100% safe environment is a bit optimistic?

A reasonable hope would be that in performing macro-dimension laboratory experiments one would not be exposed to any additional new dangers which one would not otherwise be subject to.

The good news is that you will not be exposed to any new dangers which you would not otherwise be subject to – at least not in the physical sense.

However, we cannot promise that your scope of understanding and/or perception will not be expanded.

Some folks find it unsettling to see the world without the consensus reality glasses with which we each are outfitted over the course of our up-raising.

If you have no intention of putting in the work to better your situation, perhaps it would be more comforting to remain unaware of the true situation until the moment of inevitability strikes.

But, if you happen to be inclined toward enabling yourself to do something about your situation, then you would be advised to open your eyes and take note of your situation whether it happens to be unsettling or not.

Many of the experiments in the Macro-Dimension Laboratory Series are of the "eye opening" variety. And many, of the experiments in this series are of the "enabling yourself to do something about your situation" variety.

We do try to present these experiments in as safe a fashion as possible. But, let's face it, these exercises are designed to point out the cracks in the cosmic egg and provide admission to the other side of the veil.

We expect more than just common sense. We expect the same degree of care and attentiveness to caution that any chemistry or physics laboratory would expect.

If you can muster that level of self-initiated care, then these experiments should be relatively safe.

Notes: Diffused Vision

Diffused vision is often misunderstood as fuzzy or out of focus. That is not the point at all.

When using your eyes "to look" there are two separate processes involved. One process involves optical focus, the other process involves the distribution of attention within the visual field.

The optical focus is associated with clarity of the field of vision. In a camera, or telescope, bringing an object into focus involves adjusting the lens so that the image is brought into clarity. When the object is out of focus it is blurry and detail is lost. When an object is in focus, it is clear and rich with detail.

Just so you know, "diffused vision" is not an associate of optical focus. Diffused vision is not achieved by allowing the eyes to go fuzzy.

Diffused vision relates to the distribution of attention within the visual field.

If your attention is gathered and assigned to a small area in the center of your visual field, this is not diffused vision. It is the opposite.

In diffused vision one will smear, spread, and/or distribute attention equally throughout the whole of the visual field.

Normally when we study an object we place as much attention on that object as possible – ignoring the rest of the visual field.

This is not the only option. It is possible to place a portion of your attention on the one object and maintain attention on other elements in the visual field as well.

A baseball pitcher will do this – as least the good ones will. When a moderate pitcher throws the ball, they will have a great deal of attention focused on the batter and catcher. However, a great pitcher will have attention on the batter, the catcher, his team, base runners, and perhaps the news crew filming from the bleachers.

In this book, when given an instruction to use diffused vision we are suggesting that you place equal attention on every point of the visual field. In other words spread your attention equally across every part of the field of vision like soft butter on warm toast.

Whether you focus your attention on the central object of your vision or allow your attention to become diffused equally across the whole of the visual field there is still a center point. There has to be. Whether you think of your visual field as oval or rectangular there is a center point.

Often we will ask you to "make the face of your partner the visual centrum." This simply means to position your head and eyes so that your partner's face appears in the middle of the visual field. When working with someone we naturally look toward them and we naturally make them the center of our visual field.

Thus instructions about the visual centrum is just an indicator to make a particular person or object the central point in the field of vision. It is not an invitation to drop your diffused vision.

We want you to use diffused vision so that all elements of the visual field are given equal attention.

Hopefully this is somewhat clear. Not to worry if it is not crystal clear. We find that it takes a bit of actual practice (as opposed to reading about or thinking about) in order to get it.

Notes: Journaling

Jotting down experimental observations after an experiment is a good thing to do – good in many respects.

Having a habit of journaling will stimulate your attention to notice details that you might otherwise let pass by as part of the general stream of non-noteworthy experience. Basically when journaling you are affirming to yourself that your experience during the experiment is noteworthy.

Making notes in a journal also provides exercise of the skills required to formulate your thoughts – as if you are relating them to another – but, without actually dipping into conversation. Discussion of the experiments is not necessarily a bad thing. However, it is a good thing to process your notes without the extra complexity of adhering to social transactions and meeting expectations that either you or another places upon you during conversation.

Knowing that you are going to make journal notes after an experiment can provide a tool to divert the pesky head-brain chatter that crops up from time to time. When your head starts talking to itself you can respond something along the following manner: "Oh, good idea Mr. Headbrain, but let's leave off talking about that in our head until later when it is time to journal about the experiment."

Making notes after each experiments helps to officially declare the experiment as over. Nothing like jotting notes down in a journal after an experiment to get the message across to that the experiment is most definitely over – nothing wishy-washy about it. This kind of formalism is good for helping to close a space.

There may come a time, for who knows what reason, that you may want to refer back to your notes – perhaps to refresh yourself about your early experiences with mirror exercises. Why this would be is unknown. And, since it is unknown, you best keep fairly decent notes just in case whatever this unknown future reason is happens to be a good one. Make sense?

Thus, I think we can say that there are several fairly decent reasons in support of keeping journal notes.

That said, let's also acknowledge that there are quite a few good reasons for being conservative with the amount of time you spend journaling.

Generally speaking, if it takes you longer to journal an experiment than it took to do the experiment, you are most likely going about it the wrong way.

As a rule of thumb, don't take longer to write about an exercise than it took to do the exercise.

You may find, from time to time, that journaling becomes a significant event – more than the exercise itself. This happens. Just don't make a habit of it.

Introduction To The MDL Series

The exercises in this book serve dual purposes.

First of all, they are spiritual calisthenics engineered to develop spiritual tone and promote higher being body well-being.

In addition, the exercises function as laboratory experiments – designed to guide you, the experimenter, through a series of discoveries illuminating the nature of the macro-dimensions.

If you plan on spending any time out-of-body, dying in the near or distant future, working for the benefit of all beings everywhere, experiencing shamanic states, or in any other way leaving the narrow confines of life as a human primate, then you are going to need as much practice in the macro-dimensions as possible.

The *Macro-Dimension Laboratory Series* is designed to give you that practice.

Enjoy.

The Macro-Dimension Laboratory

The macro-dimensions is not a cute concept to be relegated to belief and cocktail party chatter. The macro-dimensions are to be experienced and worked with in a real, practical, hands-on kind of fashion. Hence, the Macro-Dimension Laboratory – a laboratory for the study of the macro-dimensions.

In a laboratory students and researchers alike will experience two types of experiments.

Experimenters will replicate iconic experiments designed to familiarize the experimenter with the existing knowledge on the subject.

The other type of experiments could be called "cutting-edge." Cutting-edge experiments dive into new areas of research on the topic – further elucidating the topic, revealing new science about the material.

Repetition of iconic experiments gives the new researcher an opportunity to develop skills pertinent to the topic and gain a fundamental knowledge of the basics.

A good laboratory teacher will guide student experimenters through a sequence of successes. This is not designed to help the student experimenter's self-esteem. This is designed to provide the opportunity to learn. Struggle, succeed, struggle, succeed, struggle, succeed, struggle is an excellent formula for penetrating a new subject. And not to worry about character building – the struggle contains plenty of set-backs and failure.

So the good laboratory teacher is not removing failure from the pattern, they are simply choosing experiments that will eventually lead to success.

The experiments found in this book and the other books found in presented the Macro-Dimension Laboratory Series are just that – experiments. They are not beliefs to be accepted. They are not barroom platitudes to be agreed with. They are not ideas to buy into. They are not badges to earn or wear. They are what they are: experiments.

As with any experiment you are expected to put the work in – to actually do the work. You are expected to follow instructions to the best of your ability, to observe, gather impressions, and ultimately to extend the experiment into new realms.

Why go through the work of bouncing billiard balls around a table in physics? You can look up Newton's Laws Of Motion in any beginning physics book. What is the point of making observations for one's self? What is the point in experiencing the phenomena one's self? What is the point of coming to one's own understand? If you have to ask these questions this series is not for you.

If you have heard about the macro-dimensions and want to do something about it in a real way, then the exercises from the books in this series might be just what you're looking for. As in everything: your mileage may vary.

What is the Macro-Dimensions Anyway?

If you have been paying any attention at all to the latest developments in physics, or watched any public entertainment during the past several years, it would be impossible for you to not know something (however small) about parallel worlds.

Well, if you take the sum total of all parallel worlds, add in volumes and volumes of shamanic lore about said parallel worlds, and incorporated something like a hybrid composed of an expert system, and personal assistant AI you would have the macro-dimensions.

If you find it easier to rap your head around parallel worlds, then substitute the phrase "parallel worlds" whenever you see the term macro-dimensions. You will be pretty much on target. Later you can pick up the subtleties missed.

Other Books by the Author

Any Game Cookbook -- Recipes for Spiritual Gaming

"In The Any Game Cookbook you'll find a bountiful buffet of spiritual exercises; a veritable smorgasbord of gaming recipes. Each recipe is designed to transform the playing of any game into a spiritual gaming experience."

Everything Other Than Chess

24 unique games that can be played with a chess set--but are definitely NOT chess.

Intended for gamers of all skill levels, from non-chess players to chess buffs, these games are an ideal way to get acquainted with different chess pieces while expanding the boundaries of gameplay.

Just Because Club: Your Personal Metaphysical Fitness Trainer

Both traditional and innovative spiritual seekers can find something of use in this training program that contains more than 100 metaphysical exercises. The powerful series of awareness experiments are for individual personal use in everyday situations and are based on a highly successful training program tested throughout North America. Both esoteric and mundane, the exercises include such tasks as going to the supermarket, sitting in an empty bathtub, and pushing hands with the ineffable. Designed to lead to altered perceptions and to create new ideas, this metaphysical program is perfect for veteran spiritual gamers, those who are seeking new experiences, or those who are simply looking for new spiritual adventures!

The Original Handbook for the Recently Deceased

Contrary to popular belief, you don't develop lots of decorum and savoir-faire just because you're dead. Au contraire, it's not so easy to handle the odd new situations that arise in the apres vie. Well, we got good news and we got bad news. The good news is that you can learn proper apres vie social graces. The bad news is that you can only learn good social habits while you're alive--which since you're dead right now doesn't do you a lot of good. Fortunately for you, this book has been written with you in mind. But be warned--this book contains great and amazing secrets meant only for the eyes of the dead.

The Original Handbook for the Recently Deceased Workbook, Manual, Practicum

A companion book to The Original Handbook for the Recently Deceased this work serves as a Workbook, Manual and Practicum of useful exercises and experiments that one can do as practical exploration of the material presented in the original text.

From The Publisher

We are pleased to be able to share this powerful work material with other seekers of the innermost path. For a current book catalog and information on further work materials, write to Gateways at the address shown below or check our website for the most current new releases.

Gateways Books & Tapes
P.O. Box 370
Nevada City, CA 95959
Phone: (530) 271-2239
(800) 869-0658

Websites:
www.gatewaysbooksandtapes.com
www.idhhb.com